CW00486030

GROUND

A MEDITATION ON ECO-ANXIETY

by

Maddie Brew
Hannah Burns
Alina Burwitz
Olivia Furness
Emrys Hough
Jade O'Shaughnessy
Ellie Sammer
Emily Thompson
Elspeth Todd

THEATRE IN THE ROUGH

Published by Theatre in the Rough in 2023
theatreintherough.com
Registered Charity No. 1133246
ISBN: 978-1-7393825-0-6

ACKNOWLEDGEMENTS

To our contributors – Maddie, Hannah, Alina, Olivia, Emrys, Jade, Ellie, Emily and Elspeth – whose creativity and generosity are at the heart of this project.

To Dinah Dossor and Joanna Smith, whose climate café sessions, kindness, and encouragement, were inspiring and uplifting.

To The Atkinson, whose project support and warm hospitality made this project possible.

To Ed Gommon from Zero Carbon Liverpool for being a brilliant advocate and assisting the project throughout.

To our funding partners, Grosvenor/The Liverpool ONE Foundation/The Community Foundation for Merseyside.

Funded by

Supported by

Eco-anxiety:

'A chronic fear of environmental doom'
American Psychological Association[1]

'A healthy response to the ecological threats we are facing'
Climate Psychology Alliance[2]

INTRODUCTION

The climate crisis is the mortal challenge of our age. From species extinction to public health disaster, the accelerating force of rising temperatures and chaotic weather patterns now threaten life on Earth as we know it.

But, in many ways, it's also the foggiest of crises. For it's everywhere and nowhere all at once. It's unimaginably vast yet impalpably small. And it happens timelessly, in the past, the present, and the world yet to be.

Today, we may find the marks of climate crisis almost everywhere. It's in the air we breathe and the food we eat. It's in the clothes we wear and the tea we drink. It's on mountain crests and ocean beds. It's in fossils from the past and in forecasts for our future.

It's perfectly reasonable, then, that a phenomenon quite so vast should feel quite so overwhelming. It's not uncommon to be overtaken by anxiety in the face of such measureless catastrophe.

It's for this reason that the climate crisis is an emotional as well as an ecological emergency. Research has shown that, globally, a majority of young people are worried about climate change. Their responses include feeling sad, anxious, angry, powerless, helpless, and guilty.[3]

What this tells us is that eco-anxiety is demonstrably real, and those who experience it are not alone.

This book is all about eco-anxiety. But whilst acknowledging the panic and unease of the first definition printed at the head of this introduction, it moves beyond it. Instead, it embraces the second definition – the idea that anxiety can also be framed as a healthy response to the environmental challenges we face.

GROUND is an exploration of this suggestion.

We invited young eco-conscious writers, artists and creatives to take a journey through their environmental angst. Provoked initially by a series of climate cafés, they were tasked with generating imaginative connections between themselves and the non-human world. For it's in the imagination that we may begin to form strange but gainful new relationships with the earth we live amidst.

This book is a collage of these connections.

Inspired by the green and blue backdrops of Sefton's long, fragmented, coastal landscape, this collection of photography, creative writing, and graphic design is an anthology of hopeful worry. Not only does it record our relationships with the non-human world, but it attempts to collaborate with it too. And in this partnership, we see grief and doubt for sure. But a sense of kinship between what Maddie Brew describes later in this volume as 'unlikely friends' shines through just as forcefully.

This book is also intended as a toolkit of sorts. We asked each contributor to answer a series of questions to guide and inspire readers who have felt the stress of eco-anxiety. If this is you, you're not alone. We hope you find the answers useful.

Finally, accompanying this publication is a film installation originally shown at The Atkinson in Southport. You can view the film and learn more about this project at theatreintherough.com/ground.

[1] Susan Clayton and others, *Mental Health and our Changing Climate: Impacts, Implications and Guidance* (Washington, DC: American Psychological Association, 2017), p.68.

[2] *The Handbook of Climate Psychology* (Climate Psychology Alliance, 2020), p.22.

[3] Caroline Hickman and others, 'Climate anxiety in children and young people and their beliefs about government responses to climate change: a global survey', *Lancet Planetary Health*, 5.12 (2021), e863-73 (p.e863).

MADDIE BREW

What makes you anxious?
My anxiety for our planet grows alongside my internal anxiety
for life. Mine is a fine line between yes and no, then and now,
now and to come, change or the lack of.

I feel as if sometimes my choices or my actions are just silk
threads hanging on for something to happen... or not.

For our planet I seek change, but isn't that what has caused
this? Can we not go back to the old, or are we beyond that
now? I suppose we are, aren't we?

My anxiety is knowing we should advance with
our technology to work alongside our planet,
but what if we get too greedy, like we have
before, what if we just

step

a little too far. What then?

My anxiety is knowing what I do, but not
having a clue.

My anxiety is knowing nothing at all, but
having to decide.

My anxiety is neither here nor there.

But it is everywhere.

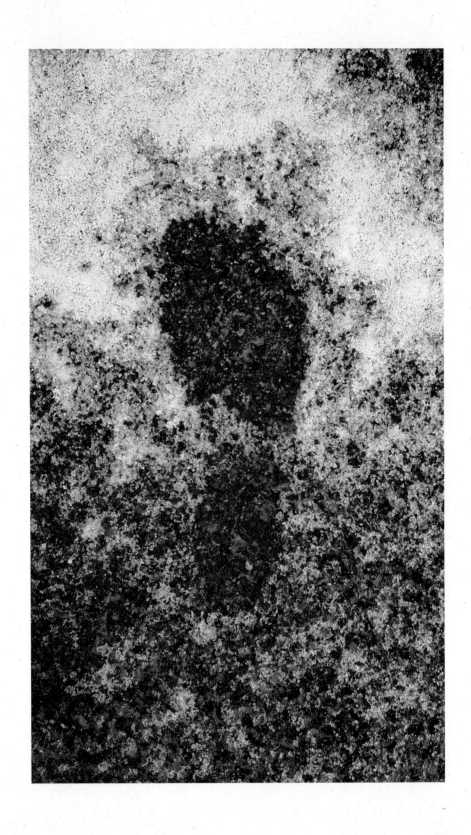

What gives you hope?

Projects like GROUND give me hope. It sparks important
conversation between unlikely friends that can be both
educational and thought provoking. We need more
uncomfortable conversations.

What would you like to save for future generations?

I'd like to save the 'quiet'.

The quiet that is never silent.

Not the kind when you sit in your house and you can hear the slight buzz from the plug sockets, or the dripping of the taps.

Sitting outside early in the morning, before the sun has woken up the rest of the world.

That kind of quiet.

Listen to it, it's beautiful.

How can we ground ourselves in uncertain times?

Climate anxiety can be hard to overcome because it's dealing with uncertainty. But I have found that looking at what we do have over what we don't helps me.

Noticing the sounds of the world, how life and nature happens around you. I like to see it as a performance sometimes, like I'm the only one there, sitting in the front row. And the world is dancing and singing just for me.

Sit on the front row of nature, observe its wonders.

TIME STAMPED

I listened to the wind.
i stood myself in silence
i had heard this story before,
but I had never listened.
to the wind, i am but a body
like a tree i stood... in silence
i heard her whispers, i felt her breathe on the back of my neck
shhhhhhhhh

i listened to the wind,
like a tree i stood in silence
she brushes through my hair to calm me, she keeps me alive
i watched her running through the winter trees
I could not see her but I watched the branches like lightning
hitting the sky.
i watched and i listened

to the wind,
i heard her talking to the birds,
guiding them, giving them directions.
she is polite, they thanked her, i thank her.

The birds wish me a good day.
i wish for guidance,
the wind takes my breath and the cold shudders my fingers
but it will not stop me, only i can stop me.
my intentions come from the skies
as i look up to new adventures
new pathways like gaps in the clouds.
the clouds congregate to shed their skin
to feed the seeds of new ideas.
my seeds are only just finding first roots,
their own pathways through the ground
and beginning to sprout.
the sky shall shed its skin
and the birds will guide me to the sun.
the cold may stunt my leaves but that won't stop me
I am an evergreen.
listening to the wind.

HANNAH BURNS

What makes you anxious?
Time.

What gives you hope?
Students. People that chose to learn every day. We're all students of the earth, learning from it and growing with it.

What would you like to save for future generations?
The freedom to touch and breathe and live with nature.

How can we ground ourselves in uncertain times?
Spend some time outside, ground yourself in nature. Touch, smell, listen and see all there is around you.

ALINA BURWITZ

How can we ground ourselves in uncertain times?

Go outside. Leave your devices in your pocket. Be outside and pay attention to something outside of yourself. Remember it's not just you: it's never been just you.

Have big conversations in little chunks. It doesn't need to be all in one go. Without joy, without enjoyment, none of this means anything.

So, remember the small reasons why you care. Like how spring comes slowly then all at once. Like how water passes over rock until it has made its own path. You're not alone in this.

Fresh air?
Bottle it.

What makes you anxious? People not doing the small stuff, let alone the big stuff. Hot summers getting hotter and everyone celebrating it, winters getting warmer, and people not questioning how these things are connected.

What gives you hope? How simple the small stuff is. How resilient the habitat is.

What would you like to save for future generations? Seasons. Safety. Seeds.

OLIVIA FURNESS

What makes you anxious?
Ignorance.

What gives you hope?
Future generations.

What would you like to save for future generations?
Plants and animals.

How can we ground ourselves in uncertain times?
Be proactive – do whatever you can to help.

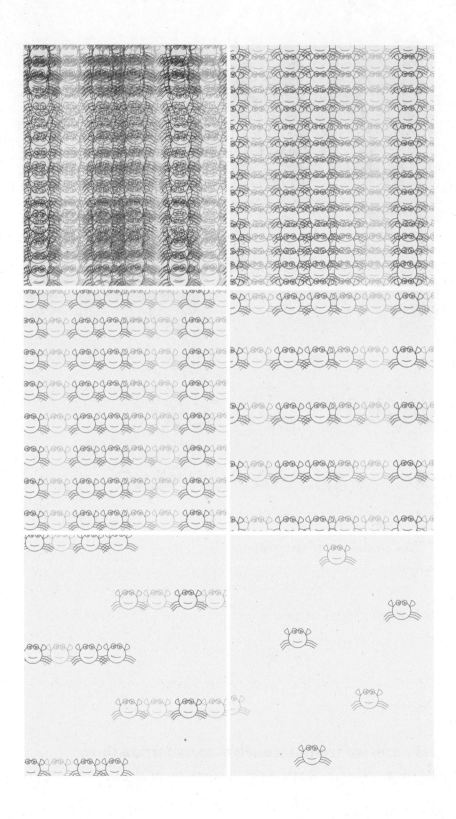

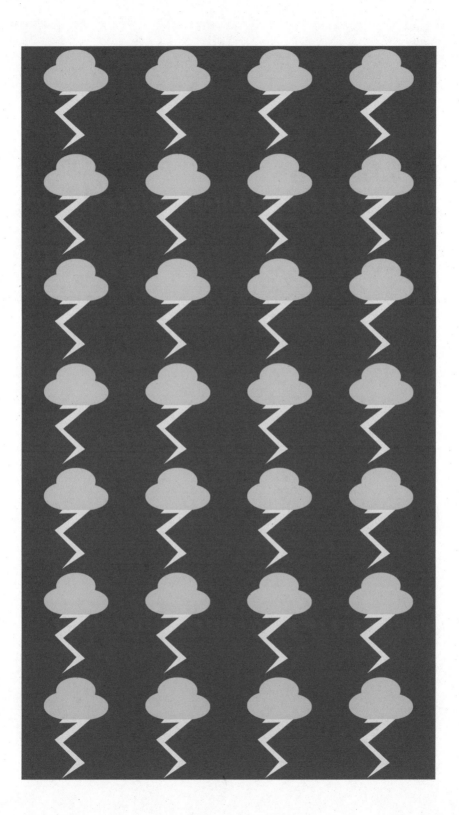

EMRYS HOUGH

Why is your head
in the sand?

TIME IS RUNNING OUT.

What makes you anxious?

That we are replacing life for the material. Trees for fences, fish for bottles, birds for planes. What if we were born 100 years later, would we still be able to experience all nature has to offer?

What gives you hope?

The people that care, the amount there are, the spirit and heart that we use to fight for this home that is ours.

What would you like to save for future generations?

Wilderness and wonder, the great outdoors and all it has to offer.

How can we ground ourselves in uncertain times?

To take action, anxiety is released through movement, do something be it creative or in-field, minute or far reaching. Release it physically, take a step in the right direction.

JADE O'SHAUGHNESSY

What makes you anxious?
Forgetting and letting the world decay
Another problem to deal with on another plate.
When the news is just another chore
And I'm scared and I'm tired and I'm just a bit bored
Stick my fingers in my ears, shout louder than the waves
Until the day comes that I can't shout again.

When I'm silent, and another has my say
Will I feel I've made the world a better place?

What gives you hope?
The thought of my generation growing up, showing up.
Putting value in the leaves and trees and air we breathe.

How can we ground ourselves in uncertain times?
I would say focussing on the small changes and small effects
that we can have as individuals, like signing petitions, going
to protests, holding local politicians accountable. It can seem
like such a huge thing but if we all play a part in trying to bring
about change then I believe that we can.

What would you like to save for future generations?
Ever since I was a kid, I used to go to the beach and pick up
seashells with my family. We would walk through the woods
with our dog, make a day of it, and when we got home we
would clean the seashells and put them in this huge glass vase
we had in the bathroom.

In 2021, not long after I had met my now girlfriend, I took
her to the beach and we collected shells together, and walked
through the woods together, like I did with my family when I
was a kid.

There is something so magic about these places, the beaches
and the pinewoods, they feel like the past and the present and
the future all at once.

I don't want them to lose their future. I don't want the kids of
future generations to lose them, or to lose the air they breathe
in order to be able to use them. I want to save all the future
memories that are to be made in these wonderful spaces in
nature.

ELLIE

ELLIE SAMMER

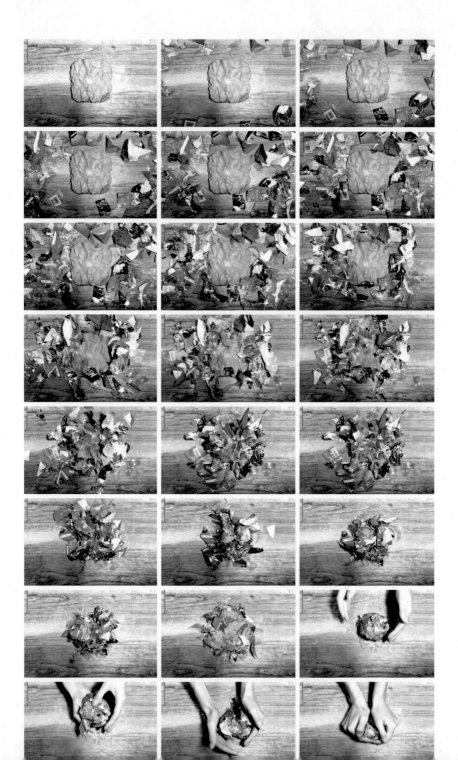

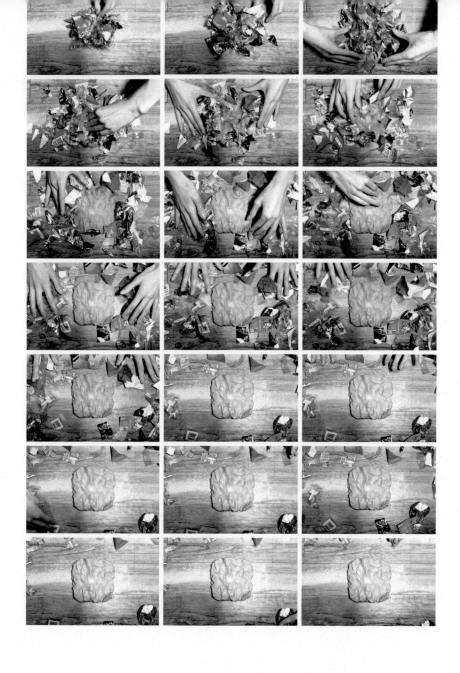

Anxiety at my fingertips

Anxiety now with every breath,
for the future of the earth
what will be left?

your jagged edges you push into
my spine as you lead me
to my decline.

my face you cannot see, so you carry
on needlessly.

our planet is suffering cant you tell?
we watch in horror, as nature is lost,
our once lush forests. now just a cost.

we can't rewind time and think what if
we must stop look around and try and
save whats left of it.

my guilt.

- plastic on all my food.
- meal deal wrapping.
- flying to visit family.

- cigarette buts from when i was a teen, that are probably still around.
- plastic in the sea I can't see.
- my flat block doesn't have a recycling system.
- not doing that beach clean up i signed up for.
- being a hypocrite.
- plastic toys from when I was a kid.
- buying new clothes from fastfashion companies, and not charity shops.
- plastic straws at bars.
- tampons with plastic applicators.
- getting taxis when I could walk there.
- not appreciating nature enough.

I know I can do better and I will.
me and my guilt.
                    Ellie Sumner.

What makes you anxious?
Not knowing, statistics, time, ignorance, disregard, people who think it's a belief, lack of education, people in power and neglect of power. Knowing I'm such a small cog in the machine, who doesn't always believe I can make a difference and who gives up.

What gives you hope?
Children and the way they see the world. Looking back at pictures from when I was younger and my physical connection with nature. They are the future and I believe they hold the key.

What would you like to save for future generations?
The animals we have left, we need to nurture them, and all
of the plants and seabeds. I don't want to see stuffed animals
in museums and tell my grandchildren they used to be on the
planet with me, but we didn't care enough and gave up on them.

How can we ground ourselves in uncertain times?
Self-care. Not in the face mask, bath-style – even though they
are both great! Just taking some time for yourself – you can't
carry the world's problems completely on your shoulders. You
are one, but when we come together, we can share the weight.

Speaking about what's bubbling away inside both helps you
and shares knowledge about the climate crisis – you're helping
yourself by educating others.

Take time out: don't always follow everything that's wrong with
the planet, you also need to enjoy what we're trying to save.

EMILY THOMPSON

What makes you anxious?
The speed at which the climate is deteriorating.

What gives you hope?
During the pandemic, the environment started to repair itself.
Dolphins started to return to Blackpool, the oceans were
clearing and no longer a murky brown colour, and this didn't
take long! If we slowed down a little bit, we could see small
changes happening.

What would you like to save for future generations?
Water is one of the most important natural resources. Save the oceans. Water is tranquil and therapeutic if you allow yourself to absorb the sights and sounds.

How can we ground ourselves in uncertain times?
Spend more time in the outdoors. Play with the natural materials that our world gifts to us. Breathe and fill your lungs with fresh air, touch the different elements of nature, look up and take in the views that you don't normally see.

Be present in the outdoor space and enjoy life.

ELSPETH TODD

What makes you anxious?
As someone who is Neurodivergent, the outside has been such
a special place to learn, and the way in which society is steering
– with the use of computers etc. – I feel as though those of the
younger generation who may have learning difficulties will not
get the same experience to learn as I did. To ignite their senses,
explore and imagine their own world, rather than something
on a screen.

What gives you hope?
My home, Scotland. I live in the city of Liverpool – however,
every season, I go to visit my family, and the world up there
changes and evolves so magically and effortlessly. It gives me
hope that there are places still full of never-failing beauty, and
a place where people can go to influence them to try and make
their home the same.

What would you like to save for future generations?
I've mentioned it above but the freedom and ability to go outside.

We are not confined to our screens – however, everyone thinks we are, with fear of missing out. Even when we're in the midst of beauty there is still a need to get it on a screen.

Yes, it is amazing to get photos, but sometimes it's even more amazing to let the memories run wild and create magical times in your head.

How can we ground ourselves in uncertain times?
Embrace what we have now, endeavour on every experience, and soak up every memory.

I always think about the Lorax with the world, and how that girl, although she had never seen a tree, knew they were gorgeous and created such a vivid painting.

So imagine how soothing it would be to know you have seen the beauties of the world, and remember whatever you see is your moment. That moment has never happened in history and will never happen again in the future. The world is doing it for you in that exact moment of time.

Embrace it.